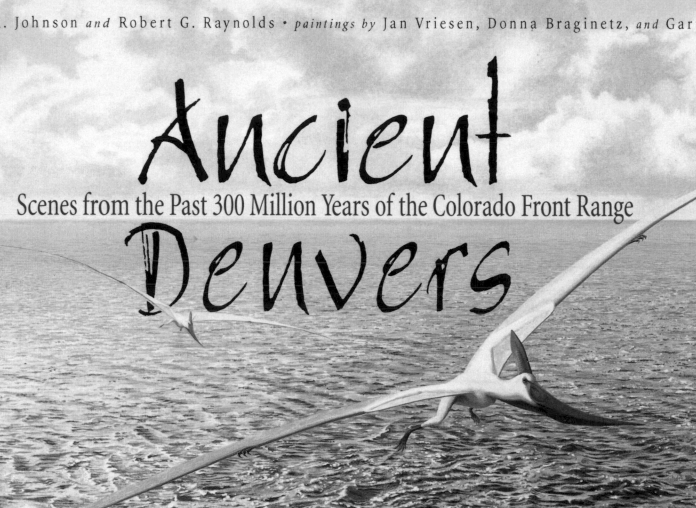

by Kirk R. Johnson *and* Robert G. Raynolds • *paintings by* Jan Vriesen, Donna Braginetz, *and* Gary Staab

Ancient
Denvers

Scenes from the Past 300 Million Years of the Colorado Front Range

Ancient Denvers

Scenes from the Past 300 Million Years of the Colorado Front Range

Rocks are found everywhere in Colorado; some are easily spied by the naked eye, while others are buried deep in the ground. From massive cliffs to jagged peaks to layer upon layer below the earth's surface, these rocks form the gateway to the past billion years of geologic history for the towns and cities east of Colorado's Front Range.

The Evidence

The prehistoric layers of rock that lie beneath our feet are hidden by a thin cover of urban sprawl, farmers' fields, roads, pine trees, and oak scrub. Many of these layers are pushed to the surface in the Front Range foothills, forming the signature scenery of the region—Boulder Flatirons, Dinosaur Ridge, Red Rocks, Roxborough State Park, and Garden of the Gods. These fantastical slabs of sandstone are the remains of ancient landscapes, different ancient Denvers.

Every rocky bluff in the Denver Basin is the exposed edge of a continuous pile of rock history. Even on Interstate 70, unsuspecting travelers zoom past one of the best displays of Colorado's past every time they head up into the mountains—the Dakota Hogback, west of Denver. The roadcut shows red, orange, gray, yellow, brown, and black layers of rock that tilt toward Denver like a row of slumped books on a shelf. The rock used to be mud and sand, *and* it used to be horizontal. Each layer represents the land surface of an ancient Denver.

When reconstructing past landscapes we use the shape of the rock layers, their composition, and fossils as building blocks. The stacks of layers may be thin or thick, some a few tens of feet thick, others more than 5,000 feet thick. By following the clues in the rock, we can read the Front Range's geologic history.

The Story

About 1 billion years ago, deep rumblings in the earth heralded the formation of large zones of melted rock that would one day become Pikes Peak Granite. This molten stew forced its way into older rocks and cooled to stone. Little is known of the next 500 million years, but by then the Pikes Peak Granite lay at the surface, granitic islands washed by a tropical sea. From 500 to 300 million years ago, the Colorado region gradually began to sink, making way for the accumulation of layer after layer of mud and lime at the bottom of a shallow tropical sea. About 300 million years ago, the sinking reversed and the granite and its surrounding rock were once again pushed to the surface and above. The overlying layers of sedimentary rocks were stripped away by erosion and the Ancestral Rocky Mountains were born.

The I-70 roadcut through the Dakota Hogback is one of the easiest places in the Denver area to see layered and tilted rocks.

Between 300 and 150 million years ago, the mountain range would slowly erode away and bury itself in its own debris. By 150 million years ago, Colorado was as flat as a pool table. For a while it was dry, but then the area began to sink again and this time the sea came from the south, flooding the region yet again. The sea waxed and waned, and finally disappeared about 69 million years ago. One more million years passed, and a series of stresses related to weaknesses in the western part of the continent caused the Front Range to lunge out of the ground, folding and breaking through its overlying blanket of sedimentary rocks. This is when the rocks that would become flatirons were tilted up from their horizontal repose.

A few million years later the dinosaurs became extinct, wiped out by a global catastrophe. Between 65 and 37 million years ago, erosion of the Rockies again buried the mountains in their own debris, and by 37 million years ago the walk from Limon to Leadville would have been a smooth, gentle stroll. Then a spectacular eruption in the vicinity of the Collegiate Range spewed fiery death across the flat landscape. Minutes after the initial explosion, a superheated cloud of molten rock surged across the low-relief terrain and buried the landscape in twenty feet of instant rock. This stony casing lay on the landscape like a steel blanket.

But everything eventually gives way to the forces of erosion, and drainages began to cut deep channels in the layer of welded ash. Sometime about 10 million years ago, the whole region began to rise up again. This time the resistant granite and other tough rocks of the mountains' core stood high while the softer rocks of the foothills and plains were more easily washed away and carried downstream to New Orleans. The flatirons were exhumed and shaped by erosion, and our present Rocky Mountain Front Range was literally unburied. Glacier-fed streams in the last 2 million years put the finishing touches on a landscape that is now called the Front Range urban corridor.

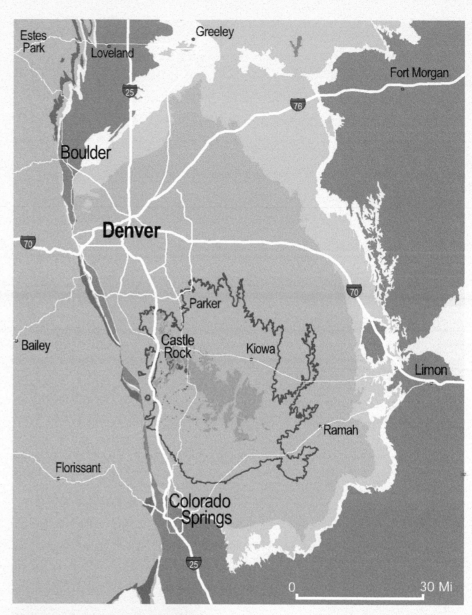

The geologic Denver Basin map shows where various rock layers are exposed at the surface in an onionlike pattern.

A sliced red onion is a good analogy for the structure of the Denver Basin.

Folding, Tilting, and Faulting—Piece of Cake!

Imagine taking a giant knife and slicing through the Front Range landscape like it was a huge piece of cake. The block diagram to the right is the cake slice, and it shows how layers of rock lie deep beneath the region's cities and parks. It also shows how the folding, tilting, and faulting of these layers bring them to the surface, exposed for all to see.

The layers of rocks are called formations, and they are usually named after the area where they were first described. Thus, the rock layer just below downtown Denver is named the Denver Formation. The column to the far right shows all of the formations that were deposited in the center of the Denver Basin.

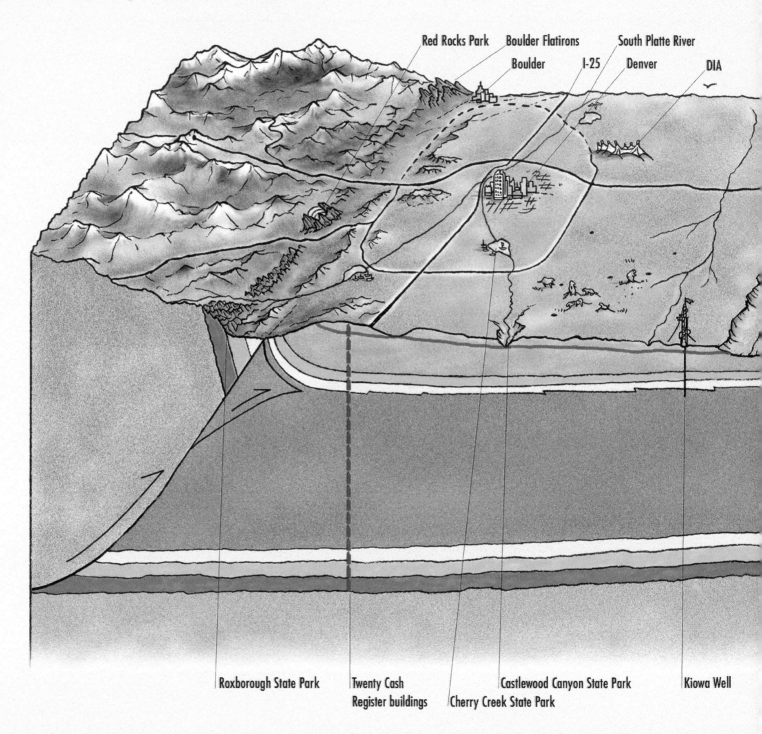

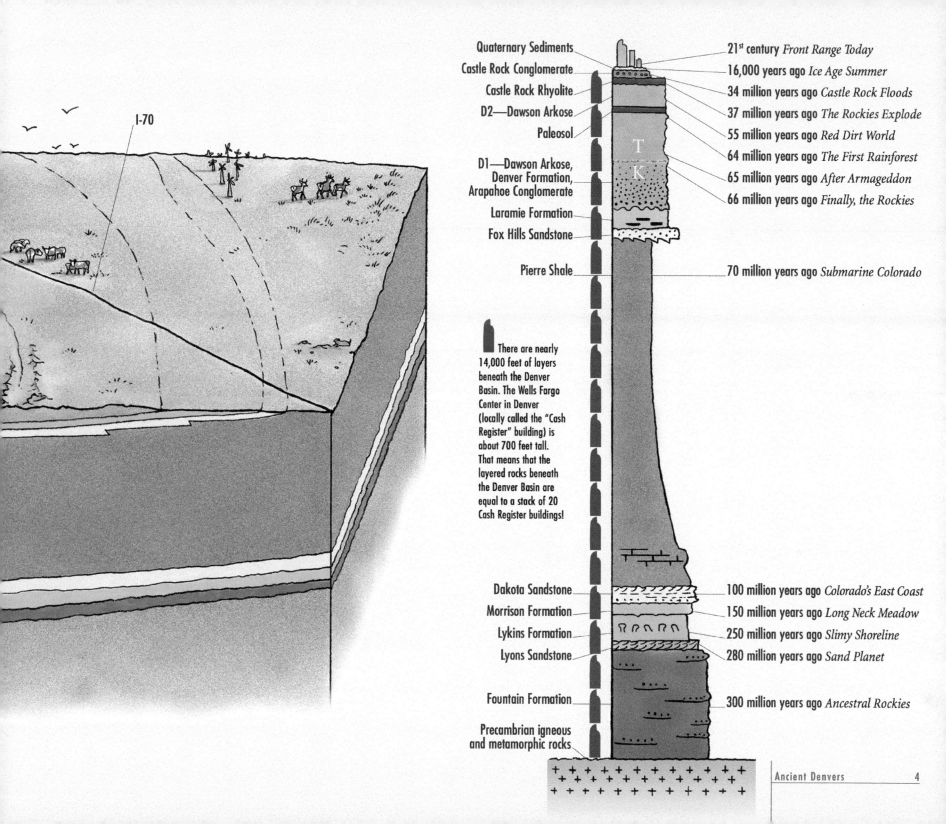

I-70

Quaternary Sediments — 21st century *Front Range Today*

Castle Rock Conglomerate — 16,000 years ago *Ice Age Summer*

Castle Rock Rhyolite — 34 million years ago *Castle Rock Floods*

D2—Dawson Arkose — 37 million years ago *The Rockies Explode*

Paleosol — 55 million years ago *Red Dirt World*

T

K

D1—Dawson Arkose, Denver Formation, Arapahoe Conglomerate

64 million years ago *The First Rainforest*

65 million years ago *After Armageddon*

66 million years ago *Finally, the Rockies*

Laramie Formation

Fox Hills Sandstone

Pierre Shale — 70 million years ago *Submarine Colorado*

There are nearly 14,000 feet of layers beneath the Denver Basin. The Wells Fargo Center in Denver (locally called the "Cash Register" building) is about 700 feet tall. That means that the layered rocks beneath the Denver Basin are equal to a stack of 20 Cash Register buildings!

Dakota Sandstone — 100 million years ago *Colorado's East Coast*

Morrison Formation — 150 million years ago *Long Neck Meadow*

Lykins Formation — 250 million years ago *Slimy Shoreline*

Lyons Sandstone — 280 million years ago *Sand Planet*

Fountain Formation — 300 million years ago *Ancestral Rockies*

Precambrian igneous and metamorphic rocks

Ancient Denvers   4

Ancestral Rockies—Fountain Formation

A giant horsetail tree fossilized in the cliffs high above Minturn, Colorado.

The Fountain Formation contains chunks of granite and quartz, remnants of a lost mountain range.

BEST VIEWING SPOTS:

Garden of the Gods Park and Visitor Center

Roxborough State Park

Red Rocks Park and Amphitheatre

Boulder Flatirons

An utterly odd Colorado lies before you, where millipedes are as large as snowboards, and dinosaurs have not yet evolved. Colorado has mountains but it also has coastlines on both its eastern and western borders. These mountains are part of a range that stretches southeast from Colorado, extending all the way to Arkansas. The surrounding forests are relatively new to the planet and are completely different from the forests to come. Gravelly streams flowing off the mountains wind through forests of 100-foot-tall scale trees and giant relatives of the horsetail rush. Buzzing insects such as dragonflies are huge; while fin-backed protomammals, reptiles, and amphibians are relatively small.

The bottom layer of the 12,000-foot-thick pile of sedimentary rock beneath Denver is the Fountain Formation. It is a 1,200-foot-thick heap of red sandstone that lies directly on top of much older, tortured melted rocks. Three hundred million years ago, the red sandstone was loose sand in the bottom of riverbeds. Now stone, the sandstone layers rest deep below Denver, but on the upturned basin margin they form the first rampart of flatirons along the Front Range. A close look at the rocks shows two big clues to their origin: (1) the cobbles of quartz and granite; and (2) the fossilized cross sections of ancient river sandbars. These features tell us that lively streams flowed from a place made of granite. Just like the granite cobble-choked streams that flow off the flanks of Pikes Peak today, so too did these ancient streams flow from a mountain made of granite. To the north, east, and west, the Fountain Formation grades into rocks that were formed beneath the sea, showing that part of Colorado was an island in a great inland sea.

Fossils aren't very common in the Fountain Formation because coarse red sandstone is abrasive and chemically harsh, so it is somewhat of a challenge, but not impossible, to reconstruct the animals and plants that lived here. At the base of the formation in Colorado Springs, fossil root systems in mudstone are direct evidence of scale trees. All parts of these trees were photosynthetic, so the trees had green trunks and green roots. By following the rocks laterally to fossil-bearing areas in central Colorado, southern Wyoming, and eastern Kansas, we are able to find additional evidence of plants and animals. Plants such as the earliest conifers, strap-leaf conifer cousins, and hugely inflated horsetails are known from the area around Vail and Minturn, Colorado. In eastern Kansas, coastline deposits are full of fossil cockroaches, millipedes, dragonflies, pole forest plants, amphibians, protomammals, and fish.

Roxborough State Park is a place where the Fountain Formation has been tipped up and exposed.

Sand Planet—Lyons Sandstone

Flagstone quarries near the town of Lyons, Colorado, are places where ancient sand dunes are dismantled to obtain building stone.

A variety of obscure, extinct animals and insects left their tracks on the ancient dunes.

It is the time of Pangaea, the supercontinent formed by the collision of all the smaller continents. The world is high and dry. Just as modern continental climates are drier and more seasonal than coastal climates, so it is on the supercontinent. Sand seas reminiscent of the Sahara cover huge stretches of the American West. These immense dunes lap up on the flanks of the Ancestral Rockies and have started to bury them. Bizarre plants and animals inhabit this giant desert—squat-bodied, square-headed protomammals the size of big dogs and a few of the world's first large-bodied plant-eaters. The plants that have adjusted to this dry and seasonal climate include some of the first conifers and cycads, and a myriad of odd seed plants that will soon be extinct.

You don't have to go far to see examples of the Lyons Sandstone, since it is commonly used to make sidewalks and fireplaces in Front Range homes. Quarried near the town of Lyons, the well-known flagstone is what remains of the ancient sand dunes that covered Colorado. The beautifully flat sheets of stone reflect the internal structure of the old dunes.

Sand dunes are formed when sand grains are blown up one side of the dune and slip down the other side. Stacks of inclined layers of sand were eventually buried and cemented to form fossilized dunes. In many places these inclined layers, known as cross beds, can be seen on the uplifted flatirons. These rocks occasionally preserved the footprints of animals and insects that slipped and slid down the surface of the dunes. The low-lying land to the south and east of the dunes supported forests of odd and extinct plants that have no living relatives. The beautiful fernlike structures commonly seen on the rocks are not fossils at all. They are dendrites, a filamentous pattern formed when a crystal of manganese oxide grows in the flat space between two layers of flagstone.

BEST VIEWING SPOTS:

Garden of the Gods Park and Visitor Center

Roxborough State Park

Area surrounding Lyons, Colorado

Hall Ranch Open Space

This Texas *Tinsleya* is an example of a now-extinct plant that grew when a sand sea covered Colorado.

Slimy Shoreline—Lykins Formation

Layered limestone mounds are what fossil stromatolites look like today.

Thin layers of mud and limestone look like phyllo dough and are the remains of algae that lived on a stinky tidal flat.

BEST VIEWING SPOTS:

Roxborough State Park

Garden of the Gods Park and Visitor Center

Red Rocks Park and Amphitheatre

This is Colorado's worst time. Erosion from the Ancestral Rockies continues to bury the mountains in their own debris and a red muddy apron rings the low hills, all that remains of the once-towering range. This grim sweltering *sabkha*, a tropical coastline much like the southern coast of the Persian Gulf today, is home to slimy mounds of algae known as stromatolites. Formed by an unlikely partnership of photosynthetic bacteria and algae, these stinky, slimy mounds inhabit the shallow and salty pools. This is not a place frequented by animals or graced by forests.

The Lykins Formation is a series of cream-colored layers of wavy limestone amid thick piles of brick red mudstone. The mudstone is so soft that it is rarely exposed at the surface, and the thin limestone layers stick out like short, white walls. Aside from the wavy beds of limestone that are the remnants of the stromatolites, fossils are extremely rare in the Lykins Formation. This was not a hospitable place to live and the red clay is evidence of intense weathering. This is a sad thing because the base of the formation was deposited just before the largest extinction in the history of Earth, the Permian-Triassic catastrophe.

The great event took place about 251 million years ago and destroyed perhaps as much as 90 percent of all marine species and a lot of land-bound life as well. This extinction remains the single greatest unsolved murder mystery in our planet's history. But the answer to this mystery is probably not going to be found in Colorado, since the conditions at this time were not good for creatures living *or* dying.

This white limestone ridge at Roxborough State Park is an example of the Lykins Formation that was mined to make the mortar that binds bricks in Denver's Larimer Square buildings.

Long Neck Meadow—Morrison Formation

A large herd of big animals moves very slowly through a waist-high tangle of ferns and cycadophytes as the morning mist begins to clear. *Apatosaurus* are the linebackers of the sauropod world of the Late Jurassic, and big boys need lots of food. While it's possible that their long necks allow them to feed high above the ground like modern giraffes, it's perhaps more likely that they saunter through thickets, letting their tiny heads sweep back and forth, mowing off succulent branch tips and leaves. This is a time before grass, before any flowering plants, a time when the planet hosts a suite of plants that will be extinct by the end of the time of dinosaurs.

Sauropods, the long-necked dinosaurs, are probably nature's greatest metabolic mystery. How did animals that were so big manage to feed themselves with mouths that were no larger, and often much smaller, than the maw of a hippopotamus? These beasts were so awesome, so evocative that they have been fodder for artists since they were first discovered near Morrison, Colorado, in 1877. One of the great ironies of paleontology is that fossil plants are not found in the same rock layers as the bones of the world's largest herbivores. This lack of direct evidence of sauropods' habitat, coupled with the desire to depict the animals rather than their environment, has resulted in common, but less than accurate, paintings of these giant beasts marching along on pounded brown earth with a line of distant conifer trees off in the background.

The footprints of the sauropods can still be seen today in the Morrison Formation at Dinosaur Ridge. It is even possible to see dinosaur bones sticking out of the rock on the side of Alameda Parkway. But, fossil plants are extremely rare at Dinosaur Ridge. Fortunately, the 400-foot-thick Morrison Formation stretches over the entire Rocky Mountain Region. Besides offering evidence that this region was then flat as a pancake, this volume of rock provides lots of places to search for fossils. Luckily, a beautiful fossil plant site in northern Montana provides the evidence of the type of plants that fed Colorado's dinosaurs.

Cladophlebis is an extinct fern that was common in the Jurassic Period.

This 80-foot-long *Diplodocus* at the Denver Museum of Nature & Science is a skinny cousin to the much bulkier *Apatosaurus*. Although more or less the same length as *Diplodocus*, *Apatosaurus* weighed almost twice as much.

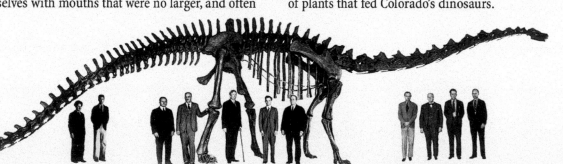

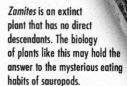

Zamites is an extinct plant that has no direct descendants. The biology of plants like this may hold the answer to the mysterious eating habits of sauropods.

100 million years ago
Colorado's East Coast—Dakota Sandstone

BEST VIEWING SPOTS

Dinosaur Ridge

Roxborough State Park

Garden of the Gods Park and Visitor Center

Interstate 70 roadcut at the Morrison exit

Deer Creek Canyon Park

On the shore of a wide, salty sea, low tide has exposed a broad-rippled sandy surface. The view along the shore shows a dank coastal forest containing herbaceous ferns, broad-leafed trees, and weird conifers. Some of the conifers grow with their roots in the edge of small tidal channels. A few *Iguanodon* dinosaurs wander down the beach, leaving their tracks on the rippled surface.

Great sheets of ripple-marked sandstone are some of the most compelling evidence that the Front Range was once a beach at the edge of a sea.

All along the Front Range, the tilted sandstone layers of the Dakota Formation show direct evidence that this sandstone was once sand on an ancient beach—ripple marks made in shallow water, muddy drapes caused by a receding tide, shrimp burrows, and the tracks of dinosaurs that left no bones. The dinosaur tracks are everywhere along the Dakota Hogback. Just think of all of the tracks one dinosaur could have made in its forty-year life. The Dakota Formation had just the right conditions for burying and preserving many millions of those tracks.

The sand is here in Colorado because a great shallow sea lay off to the east and a series of mountains hung beyond the western horizon along the far side of Utah. Mighty rivers that flowed from these mountains brought sand and gravel to the east coast of Colorado where longshore drift moved it up the coast. The Dakota Formation stretches from Utah all the way across Colorado into the middle of Kansas. In many places, fossil plants are preserved in sand, mud, or volcanic ash. For reasons dealing with chemistry, fossil bones are far less common than the footprints of the animals that lived there.

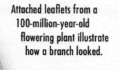

Attached leaflets from a 100-million-year-old flowering plant illustrate how a branch looked.

Fossil ferns preserved in volcanic ash unveil what the ground cover was like in the Cretaceous Period.

Submarine Colorado—Pierre Shale

BEST VIEWING SPOTS:

Rooney Road near Dinosaur Ridge

Garden of the Gods Park and Visitor Center

Fountain Creek Nature Center

Valmont Dike

The skeleton of a large pterosaur is preserved in Kansas chalk, rock that formed from seafloor ooze.

Pierre Shale fossils are often found in round rocks called concretions.

This shiny tooth from a marine lizard known as a mosasaur was collected just south of Chatfield Reservoir.

Here and there, the surface of the water is broken by the spouts of air-breathing marine reptiles the size of killer whales. In other places, schools of fish are chased to the surface by their larger fish-eat-fish cousins. It is this situation that draws the attention of the soaring pterosaurs. These reptiles fly on wings made of a filament attached to a grotesquely elongated ring finger. Looking and living like pelicans, these finger-flying reptiles are the largest animals to ever fly. Landing on the water's surface to feed is a necessity, since the nearest shoreline is hundreds of miles to the west. But landing proves dangerous in a sea full of sharks, toothy fish, and mighty marine lizards known as mosasaurs.

It is hard to imagine Colorado 600 feet beneath the salty waves of a giant sea. But it is even harder to make a better explanation for the origin of the mile-thick layer of marine mud that lies buried beneath downtown Denver. This same layer is encountered in backyard barbeque pits in Boulder and Colorado Springs, and makes up mighty Mount Garfield on the outskirts of Grand Junction. The Cretaceous Western Interior Seaway stretched from the Gulf of Mexico to the Arctic Ocean and from western Utah well into the Midwest. Colorado was literally at the bottom of the sea—not a mile high, but a tenth of a mile deep. These conditions persisted from 100 million years ago until the very end of the Cretaceous Period. The last bit of sea drained off Colorado around 69 million years ago. Remnants of the narrowing sea were still present in North Dakota until about 64 million years ago. From that time on, the center of North America has been above sea level.

Neither birds nor dinosaurs, the pterosaurs were giant flying reptiles whose closest relatives were crocodiles and dinosaurs. These were large animals, some with wingspans greater than twenty-five feet. A flying animal so large is one with very thin and hollow bones. This type of bone is easy to crush and, thus, rare to fossilize. Fortunately, the chalk beds of eastern Colorado and western Kansas were formed by very fine organic ooze that settled to the seafloor. This ooze is superb bubblepack for delicate bones. Some of the world's finest, largest, and most complete pterosaur skeletons come from chalky hills on ranches in western Kansas.

This ammonite was found in a quarry near Boulder, Colorado.

Ammonites are shelled relatives of squid and chambered nautilus.

66 million years ago
Finally, the Rockies—Denver Formation

This fossil leaf is from a plant that is related to the avocado tree.

Two banana-sized *Tyrannosaurus rex* teeth from a home owner's basement excavation in Littleton, Colorado, show the treasures that lie just beneath the surface along the Front Range.

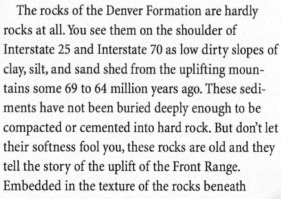

A fossil leaf excavation under a highway bridge shows the subtle outcrops of the Denver Formation.

The Colorado Front Range has formed, bringing Rocky Mountain topography back to Colorado for the first time in nearly 200 million years. The warm air and plentiful rainfall promote a bounty of plant life. The ecology of this Cretaceous forest is driven by the actions of five-ton dinosaurian plant-eaters such as *Triceratops*. The resulting vegetation is a scrubby broken forest with abundant light gaps, palm tree thickets, and patches of paddle-leafed gingers. Home to hundreds of harmless turtles, birds, lizards, tiny mammals, frogs, and snakes, this landscape is also the turf of the largest meat-eating animal ever to stand on the North American continent—*Tyrannosaurus rex*. Five tons and forty feet of bad attitude, the *T. rex* is capable of stalking and leveling animals who themselves could knock over small trees.

The rocks of the Denver Formation are hardly rocks at all. You see them on the shoulder of Interstate 25 and Interstate 70 as low dirty slopes of clay, silt, and sand shed from the uplifting mountains some 69 to 64 million years ago. These sediments have not been buried deeply enough to be compacted or cemented into hard rock. But don't let their softness fool you, these rocks are old and they tell the story of the uplift of the Front Range. Embedded in the texture of the rocks beneath Denver is the story of the formation of the mountains. Adjacent to the Front Range are layers of coarse cobbles and boulders, remnants of the rushing mountain streams that flowed off the new range into the dinosaurs' forest. Many of these cobbles are made of volcanic rock, evidence that the Front Range was clad with now-absent volcanoes. Farther from the Front Range are remains of large, sandy streams. East of Aurora and Denver International Airport, the sand and mud give way to mud and coal, the remnants of thick swamps that stank from Byers to Falcon.

Dinosaurs are relatively common in these fossil-filled rocks, and recent finds in metropolitan Denver have included parts of *Tyrannosaurus rex, Edmontosaurus, Triceratops, Pachycephalosaurus*, dromaeosaurs, and ankylosaurs, as well as many examples of the less spectacular but more numerous crocodiles, turtles, lizards, and mammals that filled out this Rocky Mountain ecosystem. Fossils are so common in this rock that almost any excavation will uncover the remains of fossil leaves, twigs, roots, and seeds.

BEST VIEWING SPOTS:

South Table Mountain
Pulpit Rock Park
Austin Bluffs Park
Palmer Park
Green Mountain

Fossil leaves are really common in the Denver Formation and can be found in almost any construction excavation. Despite being common, many of the leaves are from undescribed species.

A Denver four year old lies next to the thighbone of a *Tyrannosaurus rex*.

After Armageddon—Denver Formation

BEST VIEWING SPOTS:

South Table Mountain
Green Mountain
Highlands Ranch Open Space
Rampart Park

The landscape is lush from the warm weather and ample rain watering the forested Front Range. But this is a forest in a new world. Gone are the dinosaurs; gone are all the pterosaurs. The forests are healthy but there are only a few types of trees. Extinct relatives of sycamores, walnuts, and palms are the most common. Small nocturnal mammals roam the forest floor but none are much larger than a raccoon. Turtles and crocodiles are the largest animals on the landscape and, surprisingly, there are no land-dwelling meat-eaters. Colorado is boring, but safe.

In a flash, about 65.5 million years ago, an asteroid about the size of Denver struck Earth in the shallow seas that covered Mexico's Yucatán Peninsula. This was the largest known extraterrestrial impact on Earth and the results were devastating all around the world. Dinosaurs and their ecosystems were literally blown away. North America, due to its proximity to ground zero, was hardest hit but forests on the other side of the world in New Zealand were also flattened and incinerated. It was a humongous disaster, yet there were survivors such as certain plants, small animals and birds, and aquatic creatures. These were the lonely colonists of the new ecosystems that formed in the years and millennia after the impact. The fact that Colorado had topography in the form of the Front Range was a good thing, as survivorship here was greater than it was farther north where there were no mountains to provide protection from the blast.

This was also a time of great volcanism in the Rockies. It was during this time that the minerals of the Colorado Mineral Belt were emplaced by the percolation of gold- and silver-laced hot water associated with volcanic activity. Lava eruptions near Golden flowed out across the landscape and remain today as the mesa-capping layers at the top of North and South Table Mountains. More distant volcanoes, behaving like Mount Saint Helens, repeatedly blasted volcanic ash into the sky, dusting the forests. An excellent example of this was discovered during the excavation for Concourse B at Denver International Airport in 1991. Workers found a coal seam, the remains of a 65 million-year-old palm and walnut swamp. The coal seam contained forty-seven distinct ash layers, showing that the swamp had been dusted by forty-seven eruptions.

Excavation for Denver International Airport in 1991 exposed thousands of fossils.

A museum volunteer holds the base of a fossil palm frond.

Excavation for the Concourse B train station at DIA exposed a coal seam striped with volcanic ash.

Compound leaves of extinct sycamores were the most common fossils at the airport.

64 million years ago
The First Rainforest—Denver Formation

Leaves at the Castle Rock Rainforest site are so well preserved that actual leaf material is often present. This example also shows where an insect chewed off a portion of the leaf.

The very narrow tip of this leaf allowed for the leaf to rapidly drip dry after a heavy rainfall. Tips like this are common in modern tropical rainforests.

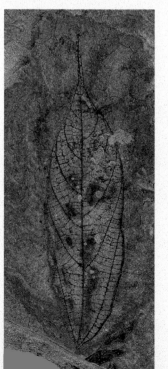

The flanks of the Rocky Mountains are covered in tropical rainforest. Tall trees dangling woody vines set the scene for a stately forest-floor cycad. It's been only 1.4 million years since the asteroid wrecked the world of the last dinosaurs, but it has been long enough for the evolution of an astonishingly diverse forest. As in modern rainforests, animals are not common. It is in forests like this that primates will soon evolve. But for now, the forest is home to a beast that is so obscure it has no common name and its scientific name is a tongue-twisting nightmare. Meet the stylinodontine taeniodont, a shuffling creature similar, perhaps, in habit to a wombat but related to the placental mammals, not the marsupials.

The most surprising fossil discovery in the Denver Basin is the Castle Rock Rainforest, found alongside Interstate 25 in 1994 by Steven Wallace, a Colorado Department of Transportation paleontologist. Located at the very top of the Denver Formation in mudstone that lay at the surface about 64.1 million years ago, this utterly unassuming site has produced more than 100 species of fossil plants. The leaves from this site share characteristics with leaves growing in modern tropical rainforests. The leaves are large in size, have smooth margins, and end in elongated drip tips that drain excess moisture from the surface. Similar sites subsequently discovered along the Interstate 25 corridor are evidence for rainforests growing on the wet, windward, eastern side of the Front Range. Today, our weather comes from the west. However, 64 million years ago, when the elevation of the Denver Basin was lower and the waters of the Gulf of Mexico closer, wet monsoons were pulled up from the gulf, drenching the Front Range with an excess of 100 inches of rain a year.

Excavation of the Castle Rock Rainforest site.

The Castle Rock Rainforest site is a very subtle eight-inch layer of sandy mudstone that lies directly on top of a fossil forest floor, complete with fossil tree roots. A small flood from an adjacent stream pushed muddy water across the forest floor, burying the leaf litter. Splitting the soft, wet mudstone reveals intact 64 million-year-old rainforest leaves that literally peel off the rock and blow away. Due to its scientific importance and its proximity to the interstate, the site is not open for visitation.

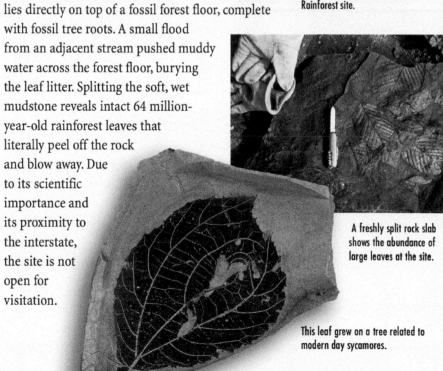

A freshly split rock slab shows the abundance of large leaves at the site.

This leaf grew on a tree related to modern day sycamores.

Red Dirt World—Paleosol-Dawson Arkose

A subtropical rainforest lines the red banks of a medium-sized river. Crocodiles and aquatic plants fill the channel, and a pair of hippolike *Coryphodon* meander along the shore. Ten million years have passed since the rainforest first formed on the flanks of the Front Range. The world is now in the grip of an intense phase of global warming. There are no polar ice caps, and crocodiles and turtles live above the Arctic Circle.

A distinctive layer of red, purple, yellow, and orange clay rings the Denver Basin, surfacing near Parker, on the ridge above Highlands Ranch, and at Paint Mines Interpretive Park. In places, this layer is twenty feet thick. It is the remains of a soil that formed beneath a tropical forest in some of the warmest conditions Colorado has ever seen. Lying above the fossil soil (paleosol) is a thick sequence of white sandstone composed of fragments of Pikes Peak Granite. This formation is known as the Dawson Arkose and its existence is evidence of the uplift and erosion of Pikes Peak. It remains unclear how long it took for the red paleosol to form. Fossils from the Dawson Arkose are about 55 million years old. In the 9 million years that elapsed since the time of the Castle Rock Rainforest, very little sediment accumulated in the basin. Sometime during this long interval, the soil formed.

One of the most common types of fossil in the Dawson Arkose is petrified wood. It is very common around Parker where local rock hounds have given the name "Parker wood" to pieces of butterscotch-colored agatized wood. This wood is a colorful reminder of Colorado's large forests of tropical trees. Some of these trees were four feet in diameter and more than 100 feet tall.

Paint Mines Interpretive Park offers the best natural exposure of the Denver Basin paleosol. Here you can see brilliantly colored red, purple, yellow, and orange fossil dirt. Native Americans used these colored clays to make paint.

This five-lobed *Macginitiea* leaf grew on one of the most common trees of the warm Eocene world.

A bluff of Dawson Arkose, a type of sandstone made from weathered Pike's Peak Granite.

BEST VIEWING SPOTS:

Paint Mines Interpretive Park
Daniel's Park
Rock Park (Castle Rock)
Castlewood Canyon State Park

Giant petrified trees are common in the Dawson Arkose around Parker and Elizabeth, Colorado.

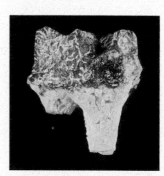

Half of a tooth is all the evidence that we have that the hippolike *Coryphodon* lived near Parker, Colorado.

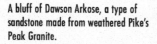

37 million years ago
The Rockies Explode—Castle Rock Rhyolite

A Capitol Hill neighborhood mansion made of Castle Rock Rhyolite.

Rhyolite blocks are commonly used to make attractive walls.

A collapsing cloud is rolling across the Colorado landscape—covering the distance between Mount Princeton and Castle Rock in as little as fifteen minutes! This cloud is bursting with superheated volcanic ash that welds as soon as it strikes the ground. The Front Range landscape is literally smothered in airborne liquid rock hot enough to melt glass. A few brontotheres are grazing near a stream, oblivious to the fact that the approaching incandescent cloud of molten ash will momentarily roast them.

For more than 15 million years, the Rocky Mountain landscape was worn flat by erosion as the uplifted Rockies shed sediment into the Denver Basin. By 36.7 million years ago, the landscape was a gentle east-dipping slope. A major volcanic eruption in the Collegiate Range catastrophically deposited a regionally extensive welded tuff across the Denver Basin.

Today this eruption is represented by a layer of rock known both as the Wall Mountain Tuff and the Castle Rock Rhyolite. In places,

Skilled quarry workers use simple hammers and chisels to make square building blocks from the welded tuff.

this layer is about twenty feet thick and it occurs on the top of buttes between Castle Rock and Monument Hill. For more than a century, it has been mined as an attractive building stone. Many of the classic Capitol Hill neighborhood mansions in Denver, including the Molly Brown House, are fashioned from rough-hewn blocks of this volcanic rock. Aside from old mansions, the rock is difficult to observe because most of the butte tops around Castle Rock that contain the rock are privately owned. However, chunks of the Wall Mountain Tuff can be seen as boulders at the base of Castle Rock, and in exposures of the Castle Rock Conglomerate at Castlewood Canyon State Park.

BEST VIEWING SPOTS:

Rock Park (Castle Rock)

Mesa tops between Castle Rock and Monument Hill

Molly Brown House Museum

Castlewood Canyon State Park

The remaining natural outcrops of the Castle Rock Rhyolite occur on the tops of certain mesas in Douglas County.

Castle Rock Floods—Castle Rock Conglomerate

Palm trees sway in a shallow canyon cut into the recently deposited volcanic rock. The eruption of the Castle Rock Rhyolite was the equivalent of pouring twenty feet of concrete on this landscape, and it took some time for the vegetation to recolonize and for the natural drainages to reassert themselves. In response to intermittent heavy rains on these rearranged rivers, catastrophic floods roar down canyons cut into the volcanic rock. These floods often sweep up unfortunate rhinoceroses and titanotheres.

The Castle Rock Conglomerate is what remains of the catastrophic floods that rushed down the canyons of the Castle Rock Rhyolite. If you climb to the base of Castle Rock, you will see refrigerator-sized angular chunks of Castle Rock Rhyolite, which had fallen into the rushing torrents. In a nifty bit of topographic reversal, Castle Rock, now the highest point around, was once the bottom of a stream channel, which was the lowest place around.

Remnants of this layer form the mesa country between Denver and Colorado Springs, extending east past the town of Elbert. In places, the conglomerate contains bones of animals that were washed into the torrential streams. The front half of a titanothere skull was collected on the grounds of the Boy Scout camp near Elbert in 1996. The fossil beds found in Florissant Fossil Beds National Monument, with their spectacular *Sequoia* trees, are the same age as the Castle Rock Conglomerate.

BEST VIEWING SPOTS:

Rock Park (Castle Rock)
Castlewood Canyon State Park

The lower slopes of Castle Rock are formed of Dawson Arkose and only the rectangular chunk at the top is Castle Rock Conglomerate.

The Castle Rock Conglomerate contains huge chunks of Castle Rock Rhyolite, direct evidence that it formed after the rhyolite was deposited.

Titanotheres were giant browsers that became extinct shortly after the time of the Castle Rock floods.

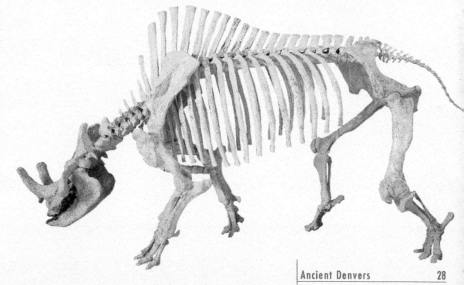

Ice Age Summer—Quaternary Sediments

An eagle-eyed backhoe operator spotted the tip of a mammoth tusk while he was digging a soil-test trench in Littleton, Colorado.

It's a summer day near the end of the Ice Age. Mammoths and camels, searching for their dinners, are wandering in and around pine trees and prairie grass. Even in the coldest of times there is still a difference between summer and winter. This summer scene of 16,000 years ago shows the landscape of what will eventually become Highlands Ranch.

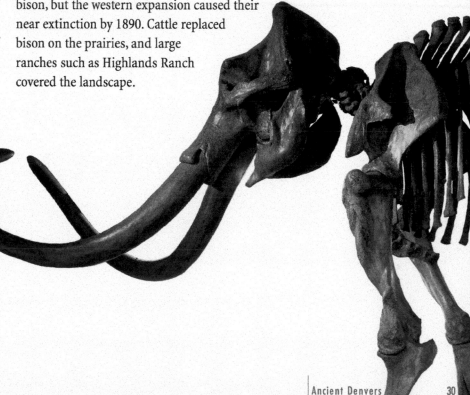

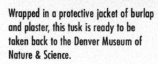

Wrapped in a protective jacket of burlap and plaster, this tusk is ready to be taken back to the Denver Museum of Nature & Science.

Between the deposition of the Castle Rock Conglomerate and the Ice Age, Colorado was uplifted to its present elevation. This increased elevation and increasingly poor weather caused rapid erosion of the soft sediments of the prairie and foothills. Areas along the Front Range were eroded by water flowing down Plum and Monument Creeks. The mesas around Castle Rock were carved during this time. In places, massive chunks of rock were washed miles from the mountains by powerful streams. The Rocky Mountains were literally being unburied and exposed.

The ice ages waxed and waned for nearly 2 million years. Glaciers formed in the mountains but did not extend out onto the plains. During this time, mammoths, mastodon, camels, horses, bison, lions, cheetahs, and giant ground sloths lived along the Front Range. Around 11,000 years ago, the first humans settled in the region and hunted bison and mammoths. The extinction of all of these large mammals, with the exception of bison, happened around this time. Bison survived and multiplied into massive herds. Horses were reintroduced to Colorado in the 1500s by Spanish explorers. Cheyenne, Kiowa, and Arapahoe used horses to hunt bison, but the western expansion caused their near extinction by 1890. Cattle replaced bison on the prairies, and large ranches such as Highlands Ranch covered the landscape.

Mammoths, like this giant from Nebraska, were the largest animals to walk on North American soil after the extinction of the dinosaurs.

Front Range Today

This rig was used to drill and retrieve a continuous, 2.5-inch-diameter core of rock from the center of the Denver Basin, near Kiowa, Colorado.

A core sample from the well at Kiowa provides a trove of scientific information about the deep history of Colorado.

Standing in a sea of homes, it's hard to believe that Highlands Ranch was actually once a ranch. Today it is home to more than 70,000 people on the southern edge of metropolitan Denver. Colorado's climate, economic opportunity, and access to outdoor recreation have caused a boom in population growth. Unlike most of *Ancient Denvers'* images, this one does not show running or standing water. High and dry, the Front Range receives only 12 to 14 inches of rain per year.

Extensive construction in the Front Range urban corridor is beginning to overprint the natural landscape. Denver's population increased 23 percent between 1990 and 2000. This rate of increase impacts the sustainability of natural resources such as groundwater. The residents of Denver drink water that falls as snow on the west side of the Continental Divide, which is then pumped from reservoirs through a series of tunnels under the crest of the Rockies, and into Denver.

Many of the suburbs of Denver and Colorado Springs do not have access to this water and must drill for groundwater. The Denver Basin is underlain by several bedrock aquifers, layers of saturated sand. These aquifers are the remains of ancient buried landscapes, and the saturated sands are actually old sand-choked streambeds and beaches. Some of the groundwater near Parker, which is pumped from depths as great as 2,500 feet, comes to the surface at nearly 85 degrees Fahrenheit, heated by radioactivity from the inner portions of the earth. This is the same water that fell as rain tens of thousands or perhaps millions of years ago. It may even have fallen on the backs of mammoths or rhinoceroses or dripped from leaves in Colorado's ancient rainforest. Today, this warm, vintage water is the key resource that allows for the settlement of large numbers of people on an arid, high plains landscape.

In 1999, the Denver Museum of Nature & Science drilled a 2,256-foot-deep well in Kiowa, Colorado, which is near the center of the Denver Basin. The hole was drilled to better understand the sequence and age of the extinct Colorado landscapes and to understand the nature of the groundwater resource and the rocks that host the water. The core from the well provided samples that were subjected to a battery of tests that dated the layers and tested their capacity to hold groundwater. Results from these tests show that the aquifers in the center part of the basin hold less water than the aquifers on the western margin of the basin. The water-bearing layers were formed during the uplift of the Front Range and the sediment near the mountains is coarse, thus able to hold more water than the finer sediment deposited farther away from the Front Range. The shape of ancient and extinct landscapes controls the distribution of the water that is so precious to the growing population of Colorado's high plains.

On the Road—
How To Use this Book

The outcrops and rock layers that piece together the amazing stories of *Ancient Denvers* are exposed in dozens of parks and open space areas along the Front Range. Outfitted with this book, a visit to any one of the following parks will open up doors to extinct landscapes visible to the naked eye. Please make sure to obey the rules and regulations of the parks. It is illegal to collect fossils and rocks within the parks, and many parks have strict rules about staying on the trails.

The following list pairs the parks with the paintings and the formations. The information in parentheses matches up with the painting/formation key. The "❖" symbol indicates that there is a visitor center and/or exhibits on site. The "●" symbol indicates that the site may only be visited by appointment.

A brief description of the geology of and specific directions to each park can be found at **www.dmns.org**, just click on View Ancient Denver Landscapes!

Parks

In and around Denver
1. Barr Lake State Park❖ (K, P)
2. Bear Creek Lake Park (G–K)
3. Chatfield State Park (K, P)
4. Cherry Creek State Park (K, P)
5. Deer Creek Canyon Park (A–G)
6. Dinosaur Ridge❖ (E–G)
7. Golden Gate Canyon State Park (A)
8. Green Mountain (K)
9. I-70 roadcut at the Morrison exit (E–G)
10. Ken-Caryl Ranch (A–G)
11. Morrison Natural History Museum❖ (A–F)
12. Mount Falcon Park (A–C)
13. North and South Table Mountains (K)
14. West Bijou Site, Plains Conservation Center❖● (K)
15. Red Rocks Park and Amphitheatre❖ (A–D)
16. Sand Creek Drainage, Bluff Lake Nature Center❖ (K, P)
17. South Platte Park, Carson Nature Center❖ (K, P)
18. South Valley Park (D–F)
19. Standley Lake (K)
20. Wheat Ridge Greenbelt (K)

In Douglas County
21. Rock Park (Castle Rock) (M, O)
22. Castlewood Canyon State Park❖ (M, O)
23. Cherokee Ranch❖● (M)
24. Columbine Open Space (K, O)
25. Daniel's Park (M)
26. Greenland Open Space (K–M)
27. Highlands Ranch Open Space (K–M)
28. Nelson Ranch (A–F)
29. North Willow Creek (G–K)
30. Prairie Canyon Ranch● (M–O)
31. Roxborough State Park❖ (A–H)

In and around Colorado Springs
32. Austin Bluffs Park (K)
33. Bear Creek Nature Center❖ (B–G)
34. Paint Mines Interpretive Park● (K–M)
35. Fountain Creek Nature Center❖ (G)
36. Garden of the Gods Park and Visitor Center❖ (B–G)
37. Palmer Park (K)
38. Pulpit Rock Park (I–K)
39. Rampart Park (M)
40. Starsmore Discovery Center at North Cheyenne Cañon Park❖ (A)
41. Ute Valley Park (H–I)

In Cañon City
42. Dinosaur Depot❖ (A–G) (not on map)

In and around Boulder
43. Betasso Preserve (A)
44. Boulder Flatirons (B)
45. Contact Corner (A, B)
46. Crown Rock/Flagstaff Sill (B with an igneous intrusion)
47. Eldorado Canyon State Park (A–C)
48. Hall Ranch Open Space (A–C)
49. Heil Valley Ranch Open Space (A–F)
50. Marshall Mesa (H–I)
51. National Center for Atmospheric Research Trailhead to Mallory Cave (B–G)
52. Rabbit Mountain Open Space (E–G)
53. Settler's Park (B)
54. Valmont Dike (G with an igneous intrusion)

In and around Loveland and Fort Collins
55. Devil's Backbone❖ (E–G)
56. Horsetooth Mountain Park❖ (A–C) (not on map)
57. Lory State Park❖ (A–C) (not on map)
58. Rimrock Open Space (C–G) (not on map)

Painting/Formation Key

A. Precambrian igneous and metamorphic rocks

B. Fountain Formation (*Ancestral Rockies*)

C. Lyons Sandstone (*Sand Planet*)

D. Lykins Formation (*Slimy Shoreline*)

E. Morrison Formation (*Long Neck Meadow*)

F. Dakota Sandstone (*Colorado's East Coast*)

G. Pierre Shale (including the Benton and Niobrara Formations) (*Submarine Colorado*)

H. Fox Hills Sandstone

I. Laramie Formation

J. Arapahoe Conglomerate (D1) (*Finally, the Rockies*)

K. Denver Formation and Lower Dawson Formation (D1) (*Finally, the Rockies; After Armageddon; The First Rainforest*)

L. Denver Basin paleosol (*Red Dirt World*)

M. Upper Dawson Formation (D2) (*Red Dirt World*)

N. Castle Rock Rhyolite (also known as the Wall Mountain Tuff) (*The Rockies Explode*)

O. Castle Rock Conglomerate (*Castle Rock Floods*)

P. Quaternary Sediments (*Ice Age Summer*)